CANADA
UNITED STATES OF AMERICA
Gulf of Mexico
MEXICO
CENTRAL AMERICA
N
Where ruby-throated hummingbirds spend the summer
Where ruby-throated hummingbirds spend the winter

Ruby-throated hummingbirds are tiny—they weigh less than a nickel—but every spring they fly up to 2,000 miles from Mexico and Central America to spend the summer in the United States and Canada, where they build their nests and have their babies. In the fall, they fly all the way back again to spend the winter where it's warm.

For the staff and pupils of Highley Primary School
N. D.

For my friend Neda
J. R.

First U.S. edition 2019

Library of Congress Catalog Card Number 2018961950
ISBN 978-1-5362-0538-1

19 20 21 22 23 24 CCP 10 9 8 7 6 5 4 3 2 1

Printed in Shenzhen, Guangdong, China

This book was typeset in Aged.
The illustrations were done in watercolor and watercolor pencil with gold ink.

Candlewick Press
99 Dover Street
Somerville, Massachusetts 02144

visit us at www.candlewick.com

CANDLEWICK PRESS

Hummingbird

Nicola Davies illustrated by Jane Ray

Granny's in her garden with her granddaughter.

"Keep still," she whispers to the little girl, "and they'll come!"

The child holds her breath.

And they *do* come. . . .

Their feathers flash in the slants of light. Their wings make the sound of their name, beating fast as thought: *Tz'unun! Tz'unun!*

Tz'unun or zun-zun is the word for hummingbird in several languages used in South and Central America.

"They'll soon be gone," Granny says, "flying north like you."

The little girl looks sad, so Granny kisses her and says,

"Maybe they'll visit you in New York City."

Later, on the plane, the girl wonders how something so small could fly so far.

Down on the dark sea, a sailor has company at last. A hummingbird is sleeping in the rigging!

At dawn it wakes up and flies away, tiny and fearless, heading for the land.

Hummingbirds lose half their body weight when they fly north over the Gulf of Mexico in one long trip.

Hummingbirds feed on nectar, a sugary liquid that comes from flowers. But they need protein from insects, too, so bug dispensers give them a balanced diet.

Out on the veranda, everything is ready: the nectar feeders are filled and tiny flies buzz in the bug dispenser. Just after dawn, the hungry guests arrive for breakfast. The sisters laugh as they remember how their daddy used to say, "Hummingbirds need meat AND potatoes, same as we do!"

Spring sweeps upcountry. Flowers open: bee balm and scarlet sage, trumpet honeysuckle and cardinal flower. Insects zoom.

The hummingbirds ride the green wave, zigzagging from one pool of buzz and blossom to the next.

By April the hummingbirds are as far north as Washington, D.C., and by May they arrive in southern Canada.

STAT
LABR

A young man sets aside his schoolbooks when a hummingbird won't share flowers with a bumblebee. He laughs aloud and texts his mother a photo of the little bird, too angry for its size.

This family leaves their dinner on the table and goes outside to see hummingbirds sipping from the feeder they made out of a plastic cup and filled with sugar syrup.

Hummingbirds know exactly where they're going—and when they get there, they settle in.

The male chases other hummingbirds away so that his family doesn't have to share the nearby flowers. The female makes a nest with lichen, spider silk, and thistledown; it holds her two eggs tight, but stretches as the babies grow . . .

and grow . . .

and grow.

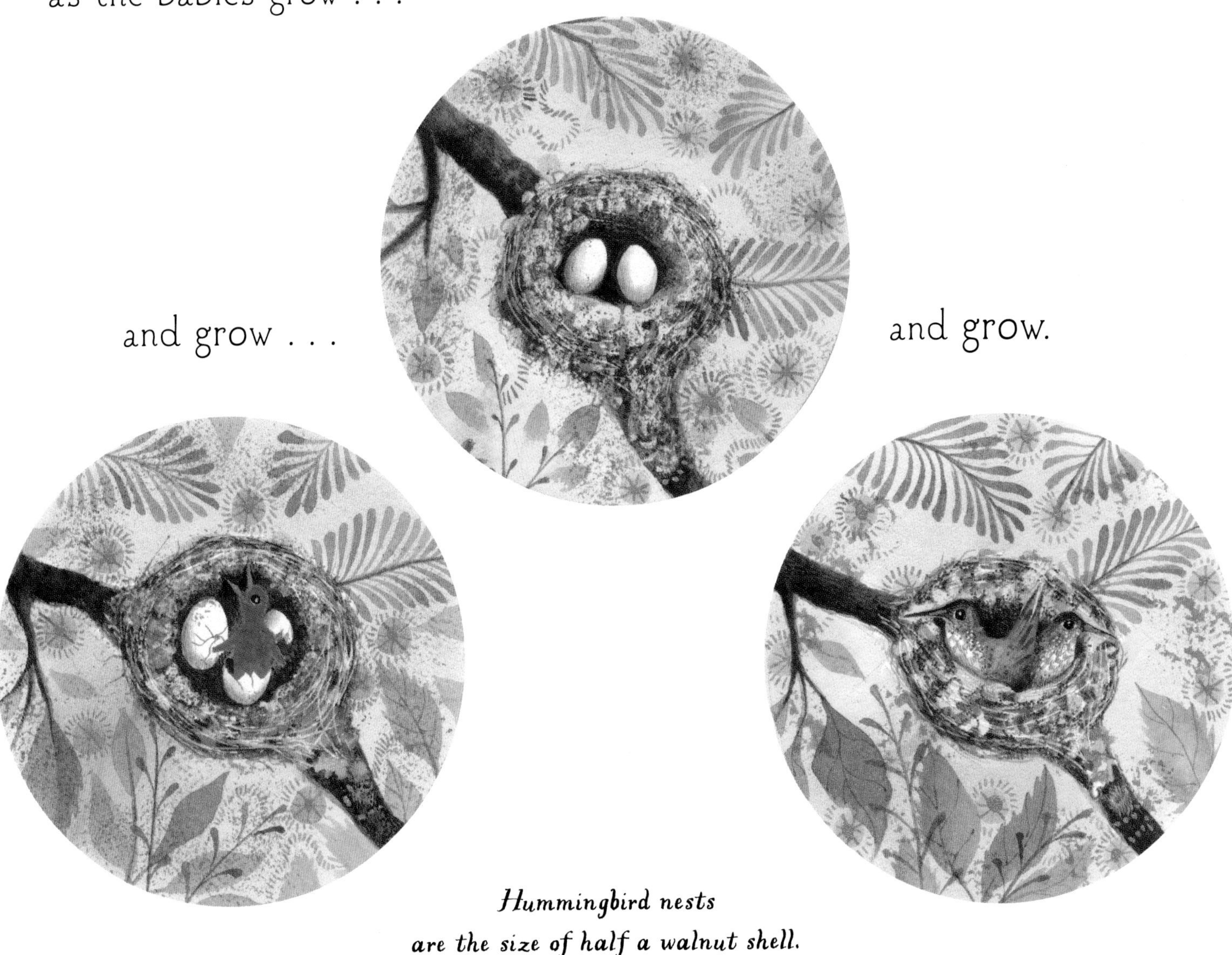

Hummingbird nests are the size of half a walnut shell.

In late summer, a little girl is walking, head down, in the park.

There's something in the grass: white, too small to cap her littlest finger. There's only one thing it could be.

Somewhere, up among the trees and green, there's been a visitor from Granny's garden!

Days are getting shorter. Soon the bugs and nectar will be nipped by frost.

Hummingbirds must fly south. The trip is long and hard for such small bodies, and many of them won't reach their destination.

Roads, houses, and cities built by humans mean that there are now fewer places for hummingbirds to refuel on their trip.

Granny's in her garden with a package in her lap—inside's a tiny eggshell wrapped in cotton balls, a letter, and a newspaper clipping.

As she reads about how hummingbirds have nested for the first time in Central Park,

there's a telltale sound around the flowers: *Tz'unun! Tz'unun!* And jewel-feathers flash in the light.

More About Ruby-Throated Hummingbirds

Ruby-throated hummingbirds are just one of more than 300 kinds of hummingbirds. They arrive in North America in the spring and leave at the end of the summer to arrive in Central America in the fall. Scientists have found out more about their migration by catching hummingbirds and putting tiny rings on their legs. Each ring carries a unique number, identifying individual birds. Every bird that is ringed is also weighed and measured, and its sex and age are recorded. When it's caught again, its ring number tells the scientists which bird it is, so they can tell how far it has traveled. These studies have shown that individual birds stick to the

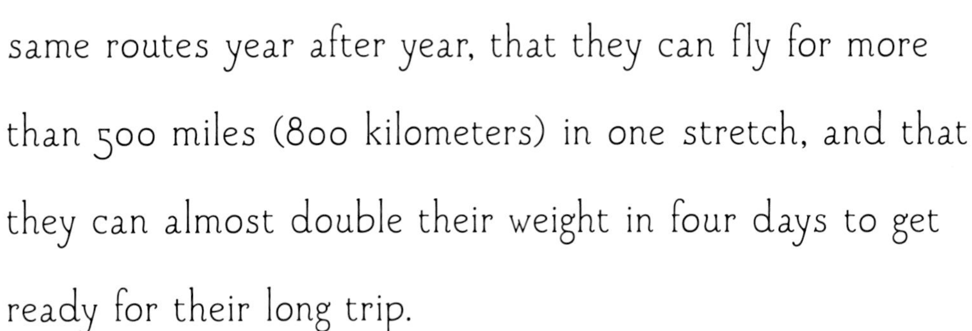

same routes year after year, that they can fly for more than 500 miles (800 kilometers) in one stretch, and that they can almost double their weight in four days to get ready for their long trip.

Catching the same ruby-throats year after year has also shown that these tiny birds can live to be nine years old!

Bibliography

Fogden, Michael, and Marianne Taylor. *Hummingbirds: A Life-Size Guide to Every Species.* New York: Harper Design, 2014.

"Ruby-throated Hummingbird." The Cornell Lab of Ornithology website. www.allaboutbirds.org/guide/Ruby-throated_Hummingbird.

Index